INSECTS: SIX-LEGGED NIGHTMARES
BOTFLIES TERRIFY!

BY M. H. SEELEY

Gareth Stevens
PUBLISHING

Please visit our website, www.garethstevens.com. For a free color catalog of all our high-quality books, call toll free 1-800-542-2595 or fax 1-877-542-2596.

Cataloging-in-Publication Data

Names: Seeley, M.H.
Title: Botflies terrify! / M.H. Seeley.
Description: New York : Gareth Stevens Publishing, 2018. | Series: Insects: six-legged nightmares | Includes index.
Identifiers: ISBN 9781538212516 (pbk.) | ISBN 9781538212530 (library bound) | ISBN 9781538212523 (6 pack)
Subjects: LCSH: Parasitic insects–Juvenile literature. | Insects as carriers of disease–Juvenile literature.
Classification: LCC QL496.12 S4535 2018 | DDC 595.7–dc23

First Edition

Published in 2018 by
Gareth Stevens Publishing
111 East 14th Street, Suite 349
New York, NY 10003

Copyright © 2018 Gareth Stevens Publishing

Designer: Laura Bowen
Editor: Ryan Nagelhout/Kate Mikoley

Photo credits: Cover, pp. 1 (botfly), 15 (top right and bottom), 19 Piotr Naskrecki/Minden Pictures/Getty Images; cover, pp. 1–24 (background) Fantom666/Shutterstock.com; cover, pp. 1–24 (black splatter) Miloje/Shutterstock.com; pp. 4–24 (text boxes) Tueris/Shutterstock.com; p. 5 Josve05a/Wikimedia Commons; p. 7 (botfly) File Upload Bot/Wikimedia Commons; p. 9 (elbow) Education Images/Universal Images Group/Getty Images; p. 9 (mosquito) Kletr/Shutterstock.com; p. 9 (tick) Henrik Larsson/Shutterstock.com; p. 11 (left) royaltystockphoto.com/Shutterstock.com; p. 11 (right) Maryann Frazier/Science Source/Getty Images; p. 13 Christian Ziegler/Minden Pictures/Getty Images; p. 15 (top left) Tacio Philip Sansonovski/Shutterstock.com; p. 17 Mark Bowler/Science Source/Getty Images; p. 21 (top) SCOTT CAMAZINE/Science Source/Getty Images, p. 21 (bottom) Thiotrix/Wikimedia Commons.

All rights reserved. No part of this book may be reproduced in any form without permission in writing from the publisher, except by a reviewer.

Printed in China

CPSIA compliance information: Batch #CW18GS: For further information contact Gareth Stevens, New York, New York at 1-800-542-2595.

CONTENTS

Bot, Warble, Heel, or Gad? 4

Central and South. 6

A Little Help from My Friends. 8

Rock-a-Bye Larva 10

Don't Feed the Botfly 12

Up and Out 14

Early Exit. 16

Looking for Love. 18

Live and Let Live. 20

Glossary. 22

For More Information 23

Index 24

Words in the glossary appear in **bold** type the first time they are used in the text.

BOT, WARBLE, HEEL, OR GAD?

Botflies go by a lot of names. Botfly, warble fly, heel fly, and gadfly are the most common. But no matter what name they're known *by*, there's one thing they're known *for*: Botfly larvae grow inside other animals—including people!

"Bot" means "larva" or "maggot," which is what we call many types of baby bugs, or insects. But these insects are special because of how scary—and painful—they are to humans. Are you ready to be grossed out by these terrifying bugs?

TERRIFYING TRUTHS

There's only one species, or kind, of botfly that lays its eggs in people: *Dermatobia hominis*, the human botfly. It's the same species that lays its eggs in monkeys!

warble fly

This is an adult botfly, but for most of its life, the botfly looks much different.

CENTRAL AND SOUTH

The human botfly is found in Mexico and Central and South America, all the way down to the northern tip of Argentina. Fully grown, these freaky fliers look a lot like a bumblebee. They're about 0.5 to 0.7 inch (12 to 18 mm) long and have a yellow face and legs. They're hairy like bumblebees, too.

But unlike bumblebees, botflies don't have a mouth because they don't need to eat! While they're still larvae, they store all the food they'll ever eat.

TERRIFYING TRUTHS

A botfly lives off the energy it stored as a larva for its whole life cycle.

A LITTLE HELP FROM MY FRIENDS

In order to get its eggs into a **host**, the human botfly needs a little help. Rather than laying her eggs directly, the female captures another bug—usually a bloodsucking bug like a tick or a mosquito—and fixes her eggs to its body with glue-like matter.

When the carrier bug lands on a host, the botfly larvae come out of the eggs. They'll live under the host's skin for 35 to 70 days. Then they'll simply fall out of the host.

TERRIFYING TRUTHS

Botfly larvae are parasitic, which means they live inside another **organism** and get all their **nutrients** from that organism.

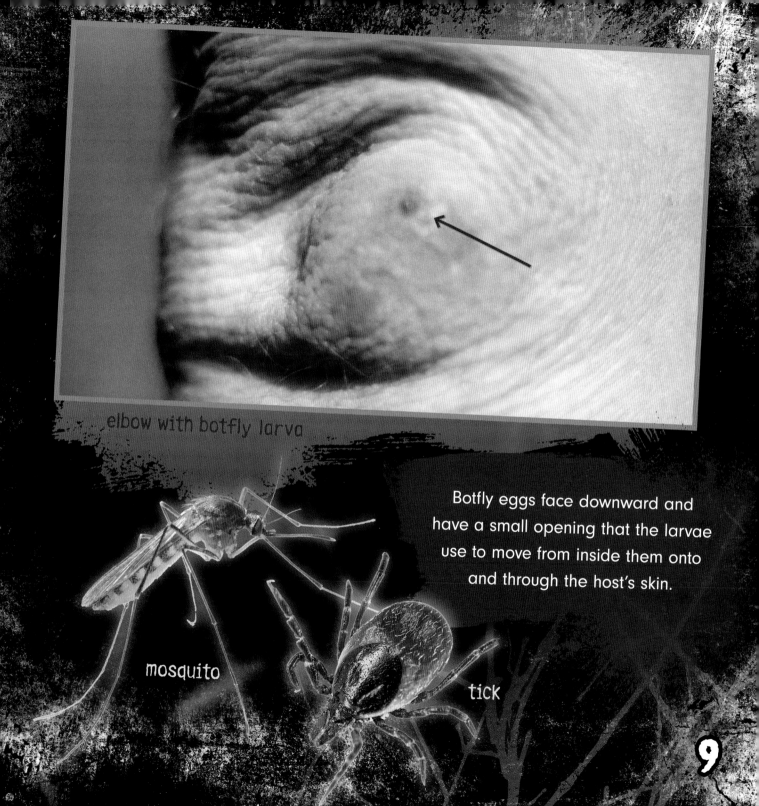

elbow with botfly larva

mosquito

tick

Botfly eggs face downward and have a small opening that the larvae use to move from inside them onto and through the host's skin.

ROCK-A-BYE LARVA

Once the larva is settled under a person's skin, a lump called a warble forms. This is why they're sometimes called warble flies. This area on the skin will often become **inflamed** and **irritated**, causing pain to their host.

The white larva is covered in pointy parts called spikes and has two small, sharp teeth. At first, it looks a little like a **tadpole,** but it then grows into a tube shape. The larva breathes through a tube that keeps the hole it entered through open.

TERRIFYING TRUTHS

Botfly larva still need to breathe when they're inside animal flesh. Sometimes a liquid will come out of their breathing hole, too. Gross!

In their last stage as a larva, botflies become round and fat.

botfly larva in a mouse

DON'T FEED THE BOTFLY

Botflies eat while they're larvae by bothering you! They eat things that fall off the skin when it's inflamed, like dead blood cells and **proteins**. To make sure they have enough to eat, larvae will even move around, making the inflammation worse so they can get more food.

Once they're adults, botflies slowly use the energy from food they ate as larvae. Botflies don't live very long. A full life cycle is usually 3 to 4 months.

TERRIFYING TRUTHS

In one case, a young girl had a botfly larva growing behind her ear—and she could hear it chewing!

The matter made by the body's response to the botfly is called exudate. This is the body trying to heal itself, but it actually feeds the botfly!

monkey infested with botfly larvae

UP AND OUT

What do you do if you find out there's a botfly larva living in you? Don't panic! It won't kill you. Once the botfly has grown to its final larval stage, it will back out of the hole it entered through, find some dirt, and **pupate**.

This is a stage of life where the botfly grows a shell to change from a larva into an adult botfly. This part of the life cycle is like what happens when a caterpillar becomes a butterfly.

TERRIFYING TRUTHS

In order to become an adult, the larva must bury itself in dirt. If the larva stays above ground, it will still grow its shell, but won't grow into an adult.

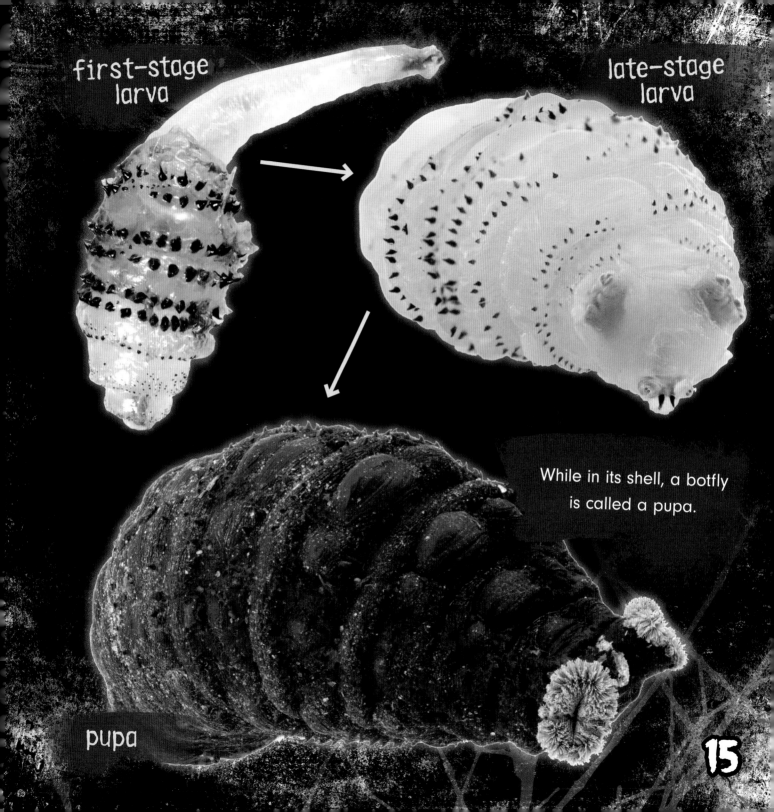

EARLY EXIT

If letting a botfly grow up and leave on its own doesn't sound good to you, botfly larvae can also be removed by hand. To do this, a doctor will cover the larva's breathing hole.

Some doctors use nail polish or **petroleum jelly**, but bacon is also a popular choice! A strip of raw bacon is put over the air hole. To breathe, the larva will crawl out far enough to be pulled out with a tool like a pair of **tweezers**.

TERRIFYING TRUTHS

You should never remove a larva by yourself—it's hard to get out. It uses the little spikes on its body to keep itself held tight inside your skin.

LOOKING FOR LOVE

Botflies don't live long once they're full-grown. A female botfly lives for about 9 days, and males live for only 4. That means they have to work fast to find a **mate**. Then, the female can find a bug to put eggs on!

This is one reason why botfly populations haven't grown outside of Mexico and Central and South America. Even if a traveler brings a botfly larva home inside them, without another botfly to mate with, the adult botfly dies without producing eggs.

TERRIFYING TRUTHS

Botflies lay many eggs at once to have a better chance of some of them reaching adulthood.

Botflies must mate within 4 days of the male hatching or the female won't be able to lay eggs!

botfly and pupa

LIVE AND LET LIVE

Botflies are pretty scary, but besides some minor pain, they won't cause you much harm. Botflies' only real goal in life is to mate, so they're not using humans as **prey**.

If you're unlucky enough to pick one up on a trip, you don't have to worry about spreading them to your town. Go see a doctor and have it removed. And maybe stop at the store on the way and pick up some bacon, just in case!

TERRIFYING TRUTHS

The longest an adult botfly can live is 9 days. Aren't you glad you're a human?

rodent botfly

deer botfly

Your family photos are much prettier than these!

GLOSSARY

host: the animal or plant where a parasite lives

inflame: to get red and usually cause pain or discomfort

irritated: red, sore, or inflamed

life cycle: the series of changes in the life of an organism, from birth to death

mate: one of two animals that come together to produce babies. Also, to come together to make babies.

nutrient: something a living thing needs to grow and stay alive

organism: a living thing

petroleum jelly: a clear, tasteless, odorless matter made from raw oil

prey: an animal that is hunted by other animals for food

protein: a necessary element found in all living things

pupate: to become a pupa

tadpole: the larva of a frog or a toad

tweezers: a small tool used to hold, remove, or pull small objects

FOR MORE INFORMATION

BOOKS

Perritano, John. *Bugs That Live on Us*. New York, NY: Marshall Cavendish Benchmark, 2009.

Romero, Libby. *Insects*. Washington, DC: National Geographic Kids, 2017.

WEBSITES

Bacon Bandages Remove Botflies
pbs.org/wgbh/nova/body/bacon-botflies.html
Watch a video about how bacon can help remove botflies from their host!

Bot Flies
wildernessclassroom.com/wilderness-library/bot-flies/
Learn all about the botfly's process of reproducing.

Wild Things: Invasion of the Bot Fly
kidzworld.com/article/580-wild-things-invasion-of-the-bot-fly
Find out about other types of botflies and where they live.

Publisher's note to educators and parents: Our editors have carefully reviewed these websites to ensure that they are suitable for students. Many websites change frequently, however, and we cannot guarantee that a site's future contents will continue to meet our high standards of quality and educational value. Be advised that students should be closely supervised whenever they access the Internet.

INDEX

adult 5, 12, 14, 18

breathing hole 10, 16

carrier bug 8

Central America 6, 7, 18

doctor 16, 20

eggs 4, 8, 9, 18, 19

exudate 13

female 18, 19

host 8, 10

human botfly 4, 6, 8

larvae 4, 6, 7, 8, 9, 10, 11, 12, 14, 16, 18

life cycle 6, 12, 14

male 18, 19

Mexico 6, 7, 18

pupa 15

shell 14, 15

skin 8, 9, 10, 12, 16

South America 6, 7, 18

species 4

spikes 10, 16

teeth 10

warble 10

warble fly 4, 10